Toroa
The Royal Albatross

Te Aorere Riddell

First published as a Māori language edition in 2001 for the Ministry of Education
by Huia Publishers. This English language edition is published in 2003 by Huia Publishers
by arrangement with the Ministry of Education.

Huia Publishers
39 Pipitea Street
PO Box 17-335, Wellington
Aotearoa New Zealand
www.huia.co.nz

ISBN 1-877283-89-4

© Text: Te Aorere Riddell 2003
© This edition: Huia Publishers 2003

All rights reserved. No part of this publication may be reproduced, stored in a retrieval system,
or transmitted in any form or by any means, electronic, mechanical, including photocopying,
recording or otherwise, without the prior permission of the copyright holder and the publisher.

National Library of New Zealand Cataloguing-in-Publication Data

Riddell, Te Aorere.
Toroa : the royal albatross / Te Aorere Riddell.
Includes index.
ISBN 1-877283-89-4
1. Royal albatross—Juvenile literature. [1. Albatrosses.]
I. Title.
598.42—dc 21

Contents

Toroa, the Royal Albatross — 4

On the Wing — 6

At Home by the Sea — 8

Feathering the Nest — 10

Feathered Friends — 11

Sea Legs? — 12

Nest Eggs — 13

A Mouth to Feed — 14

The Life Cycle of the Albatross — 15

Flight Paths and Fishing Trips — 16

Birding — 18

Laying Down the Law — 19

A Bone to Pick? — 20

Prized Plumage — 22

Bird Bones — 23

Forbidden Food — 23

The Price of Fish — 24

High Speed Collisions — 25

Foul Hooking — 26

Hatching a Plan — 28

Strikes, Storms and Stoats — 30

Index — 32

Toroa, the Royal Albatross

The sight of an albatross, the most majestic of all the ocean-going birds, is a rare privilege. Across the globe, many albatross species are facing extinction. Aotearoa New Zealand is home to more species of albatross than any other country. Here is a window into the life of the largest and best known of these birds – the Royal Albatross.

On the Wing

The Royal Albatross is the largest of all the world's sea birds. It spends most of its life at sea, making incredibly long migration flights over and around the Southern Ocean.

An adult Royal Albatross has a wingspan of between 290 and 320 cm.

The Royal Albatross spends 87% of its life at sea.

The Royal Albatross is known to fly at speeds of 115 km an hour.

At Home by the Sea

There are two species of Royal Albatross: the northern and the southern. The Northern Royal Albatross breeds at Rekohu (the Chatham Islands) and at Pukekura (Taiaroa Head) near Otepoti (Dunedin).

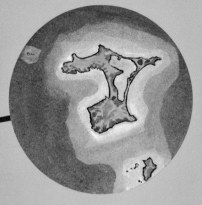

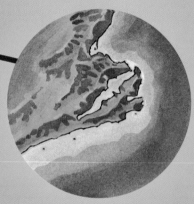

Pukekura is the only mainland albatross breeding site in the world. All other albatross nesting sites are found on islands.

Royal Albatross breeding grounds

Breeding grounds of other albatross species

Around Rekohu, the birds breed on two rocky islets called Motuhara (The Forty Fours) and Rangitutahi (The Sisters). These sites are shared with the Buller's Mollymawk, while the Chatham Mollymawk breeds nearby at Tarakoikoia (The Pyramid).

Rangitutahi

Chatham Mollymawk (Toroa Ruru)

Buller's Mollymawk (Toroa Tara)

> Every year, the albatross covers a distance of more than 190,000 km. This is equivalent to driving the entire length of New Zealand 84 times.

The Southern Royal Albatross breeds in the far south on Motu Ihupuku (Campbell Island) and Maungahuka (the Auckland Islands).

Northern Royal Albatrosses on Little Sister Island

Feathering the Nest

A Southern Royal Albatross landing at Enderby Island

The Royal Albatross has a very long breeding cycle. The time from when an egg is laid until the fledgling leaves the nest can be anywhere between 309 and 359 days. Consequently the Royal Albatross only breeds once every two years.

A bonded pair keeping company

Northern Royal Albatrosses begin returning to their breeding grounds in September. The male arrives first, followed shortly after by the female. The breeding season is the only period when the albatross spends any time on land.

After leaving the nest, the albatross doesn't return to shore until it is three to five years old. On its first few visits, the adolescent albatross spends its time familiarising itself with the breeding grounds and finding good feeding sites offshore. It is not until the albatross is between six and eleven years old that it will start to breed.

Feathered Friends

Royal Albatrosses mate for life. When looking for a partner, the Royal Albatross will gather with other unpaired birds in what is known as a gam and take part in courtship displays. A gamming albatross will raise its wings, tilt its beak to the sky, and let out a screeching 'sky call'. Other unattached albatrosses will respond, moving gradually closer and closer over a period of time, until eventually two will pair up. The pair do not start breeding until the following season. They spend a year 'keeping company' – just preening and cuddling, but not mating.

Unless its partner dies, an albatross only gams once in its life. In subsequent breeding seasons, when an albatross is reunited with its mate, it will let out a sky call in greeting, but won't perform any of the more elaborate movements associated with a gam.

Sea Legs?

Nests are made using tussock, pebbles, soil, and peat. The Royal Albatross builds a new nest every year, usually close to its old one, as it will not lay a new egg in an old nest.

When the adolescent albatross first returns to land it is very wobbly on its legs because it has usually spent at least three or more years constantly at sea.

A chicken's egg weighs 25 g on average. An average albatross egg weighs 450 g.

Nest Eggs

The Northern Royal Albatross lays its eggs between October 27 and November 27, whereas the southern royal lays between November 20 and December 20. Only one egg is laid per season. Both parents share the job of incubation; they take turns sitting on the egg for periods of four to ten days at a time. The chicks hatch after 75–82 days of incubation. Hatching is a difficult task – a chick might pound on its shell for three days before it finally emerges.

A Mouth to Feed

The hatched chick remains on its nesting site for anywhere between 230 and 280 days. Throughout this time it is fed by its parents. For the first 36 days, one parent remains constantly on guard at the chick's side. Then, the chick is left alone on the nest while both parents search for food, returning frequently to feed it.

Although not all Royal Albatross chicks survive, they do have one of the highest known survival rates of any animal. About 50% of chicks make it through to fledgling stage, and of these between 70 and 80% live to reach five years of age. This high rate is even more incredible considering that when a fledgling leaves the nest it has never flown before and must learn how to feed and how to navigate the world's oceans, all without guidance from its parents.

An albatross chick dipping its beak into the parent's mouth — the parent regurgitates food for it to swallow

The Life Cycle of the Albatross

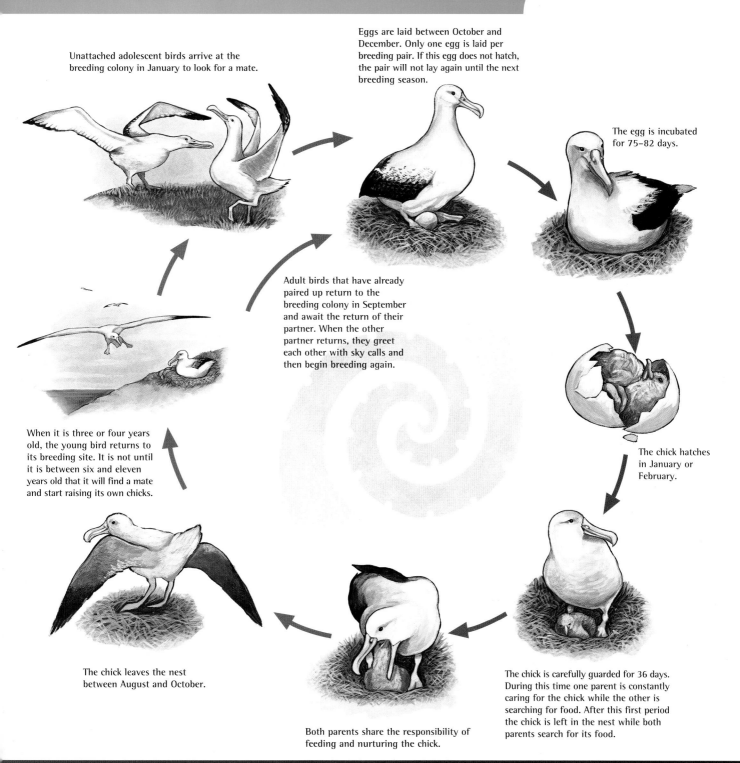

Flight Paths and Fishing Trips

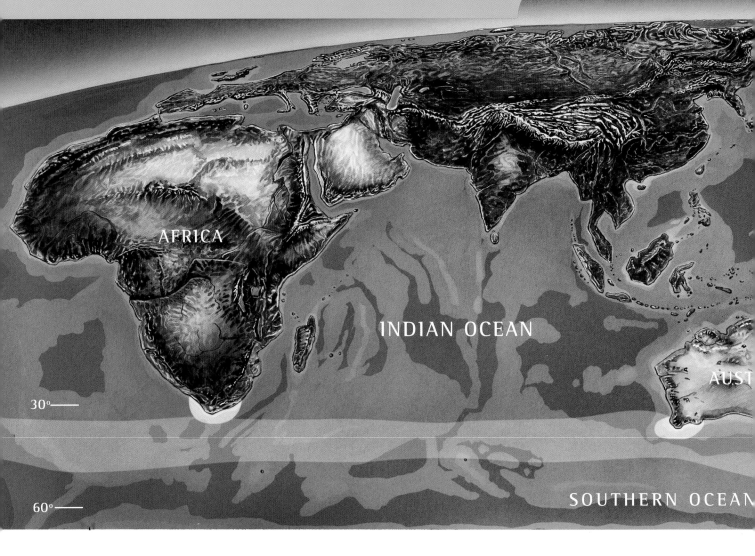

A fledgling albatross stretches its wings to strengthen them for its first flight

Northern Royal Albatross fledglings leave their nests between August and October. Adult albatrosses fly eastward, across the Pacific Ocean. They spend their winters in the Atlantic Ocean, off the coast of Argentina. From here they fly through the Southern Ocean to the south of Africa and across the Indian Ocean to Australia and Aotearoa. Generally they forage between the latitudes of 30° and 60° south, but their best feeding zones are those closer to shore, over the continental shelves.

The albatross spends long periods of time at sea in search of food. Its diet usually consists of squid and small fish.

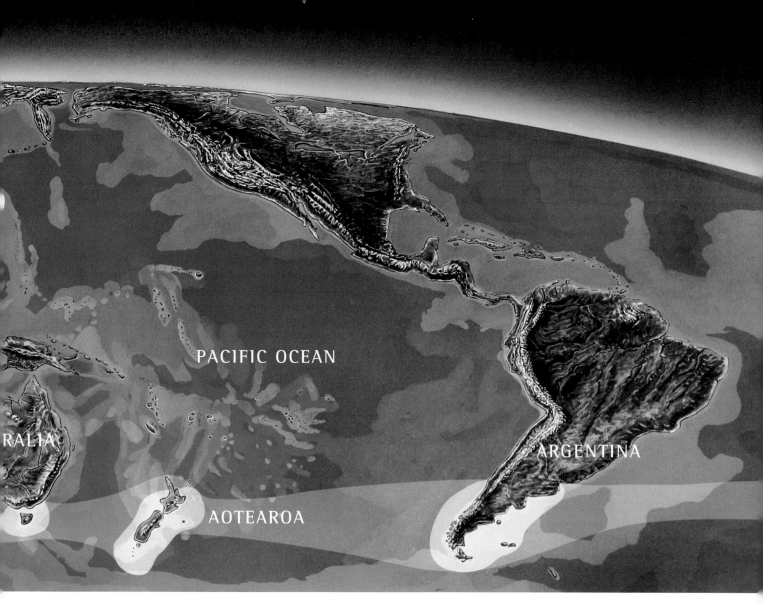

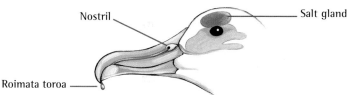

The albatross has a small gland in the top of its head which enables it to filter the salt out of the sea water it drinks so that it has a source of fresh water while at sea. The salt is excreted from the gland in droplets which run out of the bird's nostrils and down its beak, dripping off the end. These droplets are known to Māori as the roimata toroa, or tears of an albatross.

The Royal Albatross has an exceptionally long lifespan. The oldest Royal Albatross, 'Grandma', was more than 62 years old.

Birding

Albatross fledglings were a traditional food of the Moriori of Rekohu. They were also an important food source for Ngāti Mutunga and Ngāti Tama after they arrived in Rekohu (which is known to Māori as Wharekauri). Traditions concerning the gathering of albatrosses were handed down from generation to generation by the Moriori, and harvesting was carried out in accordance with these customs.

The best time to gather Royal Albatrosses was in September, just before their first flight. At this stage the fledglings are at their fattest, weighing up to 12 kg. The birds were knocked out with a club, carried to the edge of the cliff and thrown over to be collected by waiting boats.

Albatrosses from Motuhara were collected during the Second World War to be sent to New Zealand soldiers overseas

Gathering albatrosses was a dangerous occupation. Often there were rough waves, changeable winds and sudden storms to contend with. The cliffs were steep and hard to climb, and getting back to shore was often difficult. In 1900, a number of albatross harvesters died when caught in a violent storm, and harvesting ceased altogether for eleven years, not resuming until 1911.

Laying Down the Law

Some of the restrictions Moriori placed on those gathering albatrosses were that:

- metal objects were not to be used to kill the birds;

- food and alcohol weren't to be brought into the breeding grounds; and

- rubbish wasn't to be left on the islands.

Albatross oil was drunk as a remedy for coughs and colds, and rubbed on joints to ease rheumatism.

A Bone to Pick?

Albatrosses were traditionally preserved in their own fat. Processing the catch took an entire week.

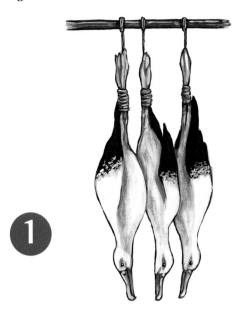

1 When the birds were brought home, they were hung upside down to prevent them from going bad.

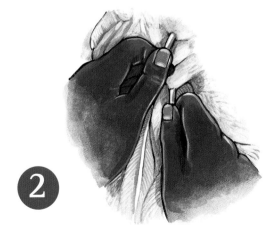

2 Next the feathers were plucked. This was a long and difficult process as each feather had to be pulled out individually.

4 The meat was then boiled in hot oil in large pots.

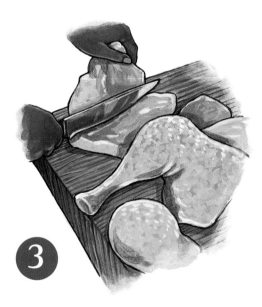

3 Once the birds had been plucked, they were gutted and the breast and thighs were cut off. The breast was considered the tastiest part to eat.

5 Once cooked, the meat was placed on racks to allow it to cool down and the fat to drip off.

6 The meat was then packed into barrels, covered in oil and sealed, preserving the albatrosses in their own fat.

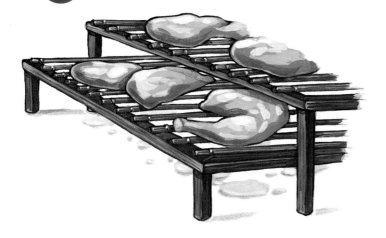

Albatrosses were sent to Parihaka as a delicacy for the prophet Te Whiti.

Moriori originally preserved albatross meat in kelp bags.

Te Niho, the wharekai at Parihaka

Prized Plumage

Albatross feathers were, and still are, highly prized. The most sought-after feathers were those plucked from the breast. These breast plumes form the distinctive adornments worn in the hair by Taranaki people. These same feathers were also used as ornaments for their ears.

In the past, albatross skin and feathers were combined with the raukawa plant to make a fragrant neck piece. Cloaks made from albatross feathers were greatly valued and worn by those of high rank. The longer inner wing feathers, when worn in the hair, were a sign of chiefly status.

Bird Bones

This pin, called an aurie by Moriori, was used to fasten cloaks

An awl, for piercing holes

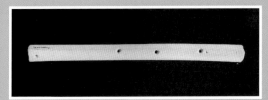
A kōauau, or flute, constructed using albatross bone

A fish hook made with albatross bone

Albatross bones were extremely useful. Some were used to make fish hooks or shaped into needles for tā moko; others were made into neck and ear pendants.

Forbidden Food

The New Zealand Government passed a law in 1921 protecting the albatross. However, sporadic albatross harvesting continued until the end of the Second World War. As late as 1941 large numbers of the birds were gathered on Motuhara to be sent overseas to the Māori Battalion. Since then, albatrosses have been gathered at intermittent periods, albeit illegally. Some Moriori and Māori groups still assert the right to harvest albatrosses for food.

The Māori Battalion on their return from the Second World War

The Price of Fish

Every year albatrosses and other seabirds are killed as a direct result of commercial fishing in New Zealand waters.

Some in the New Zealand fishing industry have developed methods to avoid catching these birds and are actively encouraging others throughout the world to do the same.

Most of the birds are either killed by long-lining boats or trawlers. Longlining is used to catch the Southern Bluefin Tuna, which fetches around US$1,500 per 100 kg. In some years, there have been up to 500 vessels fishing for Southern Bluefin Tuna in the Southern Ocean.

A typical Southern Bluefin Tuna longliner

White-capped Mollymawk

High Speed Collisions

Albatrosses following trawling vessels often end up colliding at high speed with the metal cables, or trawl warps, that haul the nets, or getting caught on sharp strands that have frayed from these main wires.

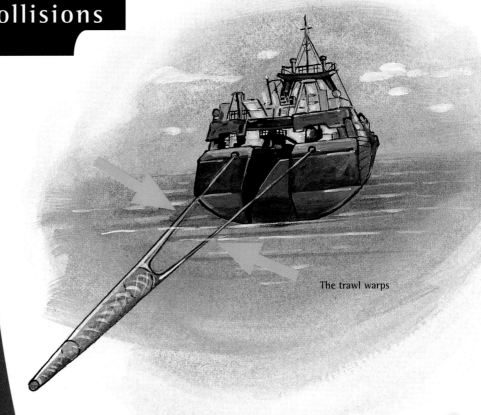
The trawl warps

Foul Hooking

Longlines consist of a main line, sometimes up to 130 km in length, with a number of shorter lines attached. These smaller lines are generally about 30 m long and each has around 3000 hooks. It takes five or six hours daily to set the longlines.

Discarded offal from fishing vessels and the baits on their lines attract large numbers of seabirds, including albatrosses and petrels. When the lines are released from the boats, the birds dive down, snatching the bait before it sinks. The bait is a deadly food: the birds can get ensnared on the large hooks and drown. There are also instances of birds being hooked through other parts of their bodies while scavenging for food near the boats. The tragedy is that if a parent bird dies so will its dependant chick. The task of rearing a chick is too great for one parent on its own.

Scientists warn that if these dangerous fishing practices continue, some albatross species will become extinct. Of the 24 species of albatross worldwide, 21 are threatened. Four, including the Royal Albatross, were listed as endangered in 2003.

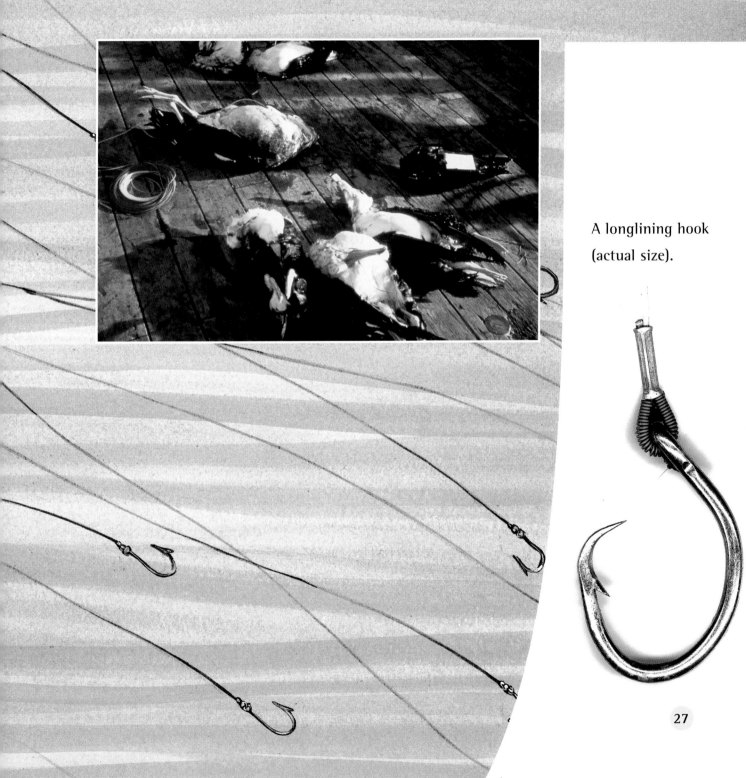

A longlining hook (actual size).

Hatching a Plan

To prevent the albatross suffering the same fate as the moa and the huia we need to act now. We can lobby the government and the fishing industry to support research and to legislate and educate to prevent the fishing practices that kill albatrosses and other seabirds.

Here are some ideas for consideration

- Streamers, or tori lines, can be fastened behind the longlining vessels to scare away the birds.

- Albatrosses tend not to forage for food on moonless or cloudy nights. Line setting could be restricted to these nights to decrease the likelihood of albatrosses being hooked.

- Thawed bait should be used. Frozen bait takes longer to sink; the longer the bait stays on the surface, the more likely it becomes that a bird will dive for it and get hooked.

- Lines should be weighted. These sink faster giving the birds less chance to snatch the bait and get hooked.

- Machines, such as the one shown below, can be used to set the lines directly into the water, instead of over the back of the boat.

- We can also support the Department of Conservation and the fishing industry in their research programmes.

A transmitter

- Gathering better information on the life cycle and habits of the albatross will allow more educated and effective conservation efforts. A great deal of information can be obtained by attaching transmitters to the birds and tracking their movements. For example, it is not known why female birds fly north in search of food, while male birds fly south. As most Southern Bluefin Tuna longlining is carried out in the northern parts of the Southern Ocean, the female birds are at greater risk than their male counterparts.

Strikes, Storms and Stoats

One of the biggest issues facing albatross chicks born at Pukekura is blowfly strike. Flies lay eggs on the backs of the young chicks, and when the maggots hatch they eat into the chicks' flesh, eventually killing them.

The chicks and eggs are also at risk from predators such as cats, ferrets and stoats. Recently, some of the albatrosses nesting at Pukekura have also been adversely affected by the higher temperatures brought on by global warming.

At Motuhara and Rangitutahi albatrosses have been producing eggs that have very thin shells and break easily. Scientists believe this is due to overcrowded conditions putting stress on the mother birds.

Raising a chick takes almost an entire year. Consequently the Royal Albatross only breeds biennially, taking a rest every second year. This means there are generally two groups of birds at any one time – those breeding and those recuperating at sea.

If a chick dies, its parents will return to breed in the following year instead of taking a year to rest. The problem with this is that if a large number of chicks die in one breeding season, there may be too many birds returning the next year to try breeding again. This creates the overcrowding that stresses the hens causing the higher number of thin-shelled eggs.

Violent storms are another problem encountered in the islets around Rekohu. Vegetation is often destroyed and topsoil is blown away leaving the albatrosses to lay their eggs on hard, rocky ground which often results in breakage.

Motuhara prior to a severe storm in 1985

Motuhara: almost devoid of vegetation some years after that storm

In order to ensure a future for the Royal Albatross we must encourage the Department of Conservation, the Ministry of Fisheries, the fishing industry, our members of parliament, the Royal Forest and Bird Protection Society, the World Wildlife Fund and all those who care for the albatross, to protect these amazing creatures now.

Index

beaks		17
bones		23
breeding	frequency of	8-15, 31
	problems with	30-31
breeding grounds		8-10
Conservation	Department of	29, 31
	laws protecting albatross	23
diet		16, 26
eggs	incubation of	13, 15
	size of	13
	hatching	13, 15
	laying	10, 12, 13, 15, 30-31
extinction		4, 27, 28
feathers		22
feeding	of chicks	14, 15
	location of feeding grounds	10, 16, 30
fishing	longlining	24-27
	safer practices	28, 29
	trawling	24, 25
flight	speed of	7
	distance covered	9
	destinations during	6, 16-17
harvesting (of albatross)	for food	18, 19, 23
	for medicinal purposes	19
	preparation methods	20-21
	timing of	18
	traditions governing	19
life cycle		15
life expectancy		17
mating		10, 11, 15
nesting		12, 14
saving the albatross		28-30
species (of albatross)		4, 6-9, 27
survival	at sea	17
	rates	14
threats (to albatross)	pests and predators	30
	from fishing	24-27
	storms	31
wingspan		7
young		13-15, 18, 26, 30

Acknowledgements

Many thanks to CJR Robertson who provided us with images, guidance and much appreciated specialist advice, also to the Museum of New Zealand Te Papa Tongarewa for supplying photographs and the Department of Conservation for supplying photographs and information. The illustrations were generously provided by Fifi Colston and the kōwhaiwhai patterns by Brian Gunson.

Photographs provided by the following:

pp. 4–5, p. 6 (top), 9 (bottom three images), 11 (bottom five images), 31 (middle) CJR Robertson, Chatham Islands
p. 9 (top) R Morris
p. 10 (top) B Dix, Enderby Island, Auckland Islands
p. 10 (bottom) MF Soper
p. 12 S Timmins, Campbell Island
p. 11 (top), 13 (bottom), 14 (top) P Moore, Campbell Island
p. 14 (bottom) CR Veitch, Taiaroa Head
p. 16 CD Roderick, Taiaroa Head
p. 24, 27, 29 (bottom left) N Brothers
p. 25 (right)
p. 29 G Sherley
p. 30 Dr N Gales, Adams Island, Auckland Islands
p. 31 (bottom), 32, A Penniket, Auckland Islands
© Crown Copyright: Department of Conservation

p. 7 (E.2961/8), p. 13 (eggs, 1.2385), p. 21 (Te Niho, B.12399), p. 22 (Parihaka family, B.12400), p. 23 (aurie, 1.1679; hole piercing implement, F.3819/2,3; fish hook, F.2767/01; kōauau, F.3486/01), Museum of New Zealand Te Papa Tongarewa

p. 6 (below) P Reese

p. 18 A Wotherspoon

p. 19 397-1/2 and p. 23 (bottom) 1663-1/4, Alexander Turnbull Library, National Library of New Zealand,
Te Puna Mātauranga o Aotearoa

p. 31 (top two images) G Murman